< GET CODING! >

Get Coding with LEGO MINDSTORMS™

By Jenna Vale

Gareth Stevens PUBLISHING

Please visit our website, www.garethstevens.com. For a free color catalog of all our high-quality books, call toll free 1-800-542-2595 or fax 1-877-542-2596.

Library of Congress Cataloging-in-Publication Data
Names: Vale, Jenna, author.
Title: Get coding with LEGO MINDSTORMS / Jenna Vale.
Description: Buffalo, New York : Gareth Stevens Publishing, [2024] | Series: Get coding! | Includes bibliographical references and index.
Identifiers: LCCN 2023006857 | ISBN 9781538288559 (library binding) | ISBN 9781538288542 (paperback) | ISBN 9781538288566 (ebook)
Subjects: LCSH: Computer programming–Juvenile literature. | LEGO Mindstorms toys–Juvenile literature.
Classification: LCC QA76.6115 .V35 2024 | DDC 005.1–dc23/eng/20230301
LC record available at https://lccn.loc.gov/2023006857

Published in 2024 by
Gareth Stevens Publishing
2544 Clinton Street
Buffalo, NY 14224

Portions of this work were originally authored by Patricia Harris and published as *Understanding Coding with LEGO MINDSTORMS*. All new material in this edition authored by Jenna Vale.

Designer: Leslie Taylor
Editor: Megan Kellerman

Photo credits: Cover and pp. 1, 7, 13, 15, 17, 19 (LEGO graphics and screenshots) the LEGO Group; cover & series art (coding background) BEST-BACKGROUNDS/Shutterstock.com; pp. 5, 7 AlesiaKan/Shutterstock.com; p. 9 wiredforlego/Flickr.com; p. 11 Vetalru/https://commons.wikimedia.org/wiki/File:Lego-mindstorms-ev3.jpg; p. 15 Priit Tammets/Flickr.com; p. 17 wavebreakmedia/Shutterstock.com; p. 21 Panasonic DC-G9/Flickr.com; p. 23 Bekahdey/ https://commons.wikimedia.org/wiki/File:Mindstorms51515.png; p. 27 icemanphotos/Shutterstock.com.

Printed in the United States of America

CPSIA compliance information: Batch #CS24GS: For further information contact Gareth Stevens at 1-800-542-2595.

CONTENTS

Building with Code.............. 4
Inside the Kit.............. 6
Coding Fundamentals.............. 8
Choosing the Software.............. 10
Starting from Home.............. 12
The MINDSTORMS Environment.............. 14
Making Connections.............. 16
Around and Around.............. 18
The Next Invention.............. 20
It's Alive!.............. 22
Virtual Remote.............. 26
Coding Today and Tomorrow.............. 28
Glossary.............. 30
For More Information.............. 31
Index.............. 32

Words in the glossary appear in **bold** the first time they are used in the text.

Building with Code

Have you ever built something using LEGO blocks? The LEGO Group's building sets have been around for many years, allowing kids to build everything from model **vehicles** to scenes from their favorite sci-fi and fantasy worlds. In 1998, the LEGO Group entered into the realm of robotics with MINDSTORMS, a set not just for building but for coding as well.

MINDSTORMS brought robotics and coding into homes at a time when personal computers and the internet were gaining popularity. For years, young builders could create and solve problems in the real world—now they could also experiment in the **virtual** world.

Many people know how much fun it is to build with LEGO blocks. Can you imagine programming your LEGO creations to interact with the real world?

Play Well

The LEGO Group is a Danish company. It's based in Denmark, a small country in Europe. The company name "LEGO" is short for the Danish phrase ***leg godt***, which means "play well." The LEGO Group has been around since 1932, but LEGO Education, the part of the company that makes MINDSTORMS, started in 1980.

Inside the Kit

MINDSTORMS kits include everything needed to build a robot, including wires, plastic blocks, beams, and gears. The picture on the next page shows the various motors and **sensors** in a MINDSTORMS EV3 kit.

The kit also contains sensors to detect touch, distance, and color. Two kinds of motors, in large and medium sizes, will be used to move your robot around. The most important part of the kit is the EV3 brick—the point of communication between you, your programs, and your robot. The EV3 brick can be powered by a couple of different kinds of battery, or power source.

Beyond the Brick

Many of the pieces used to build MINDSTORMS robots are from LEGO Technic sets. "Technic" is short for "technical." The Technic line first came out in 1977 and is focused on creating vehicles such as construction equipment and sports cars. Technic sets paved the way for MINDSTORMS by taking LEGO kits beyond interlocking bricks.

Overview

The user guide will help you identify the motors, sensors, and building blocks of your robot. All guides are also available online.

Large Motor

+ Lets you program precise and powerful robotic action.

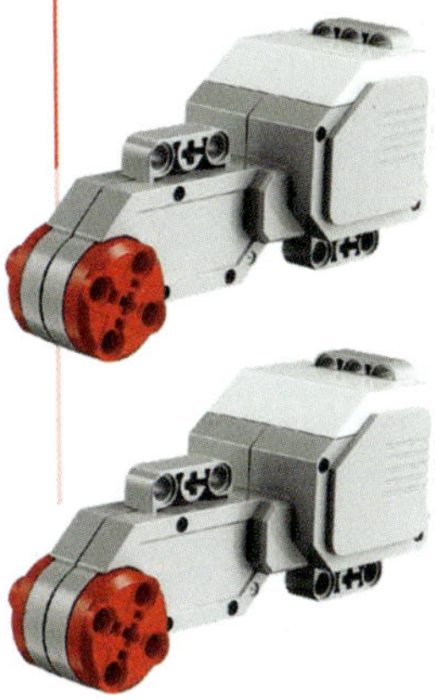

EV3 Brick

+ Serves as the control center and power station for your robot.

Touch Sensor

+ Recognizes three conditions: touched, bumped, and released.

Color Sensor

+ Recognizes seven different colors and measures light intensity.

Remote Infrared Beacon

+ Remote controls your robot from a distance.

Medium Motor

+ Maintains precision, while trading some power for compact size and faster response.

Infrared Sensor

+ Detects objects and can track and find the Remote Infrared Beacon.

Coding Fundamentals

Now that you're familiar with the hardware, or components, it's time to think about the **software**. There are a few rules about coding that you should keep in mind as you dive in.

Rule 1: Coders must know what they want the computer to do and write a plan.

Rule 2: Coders must use special words to have the computer take input, make choices, and take action.

Rule 3: Coders need to think about what tasks can be put into a group.

Rule 4: Coders need to use **logic** with AND, OR, NOT, and other logic statements as key words.

Rule 5: Coders must explore the **environment** and understand how it works.

This is Track3r from the MINDSTORMS EV3 set. To get to know it, you must first master the five coding fundamentals.

Problem-Solving

Problem-solving is one of the key skills you'll learn from writing and putting plans for your code into practice. You'll come to understand your own style of problem-solving, then you'll work to turn your plan into code the computer can understand. Sometimes computers solve problems in ways you never would have thought of!

Choosing the Software

There are many robots you can build and code with LEGO MINDSTORMS, and each kit comes with parts to build at least five different robot models. The software you use depends on which kit you're building from.

Robots from the EV3 kit, which use the EV3 brick as the computer for each robot, are **compatible** with the LEGO MINDSTORMS EV3 Home **application**, or app. Robots from the Robot Inventor kit, which use a more compact, or smaller, type of brick called the Hub, are compatible with the Robot Inventor app. Our first example will be the EV3 robot called Track3r.

The EV3 brick is the heart and brain of any MINDSTORMS robot from the EV3 kit.

The Brick and the Hub

Any robot you create with MINDSTORMS will use sensors to interact with the world around it. Sensors pick up information, or data, and send it to the EV3 brick or Hub, which then uses the information to make decisions based on the instructions you gave it.

Starting from Home

When you open the EV3 Home app, you'll choose the robot model you're going to be programming—in this case, Track3r. This is one of the easiest robots to build and program, so it will give us a very simple robot to work with.

The app will show you a selection of five missions, or sample programs, for you to view for Track3r. The mission videos will show you an action Track3r can be programmed to do. You can then either view the finished code for that mission or open a new project to try coding the activity from scratch.

This is the home page for the EV3 Home app. Track3r is all the way to the left.

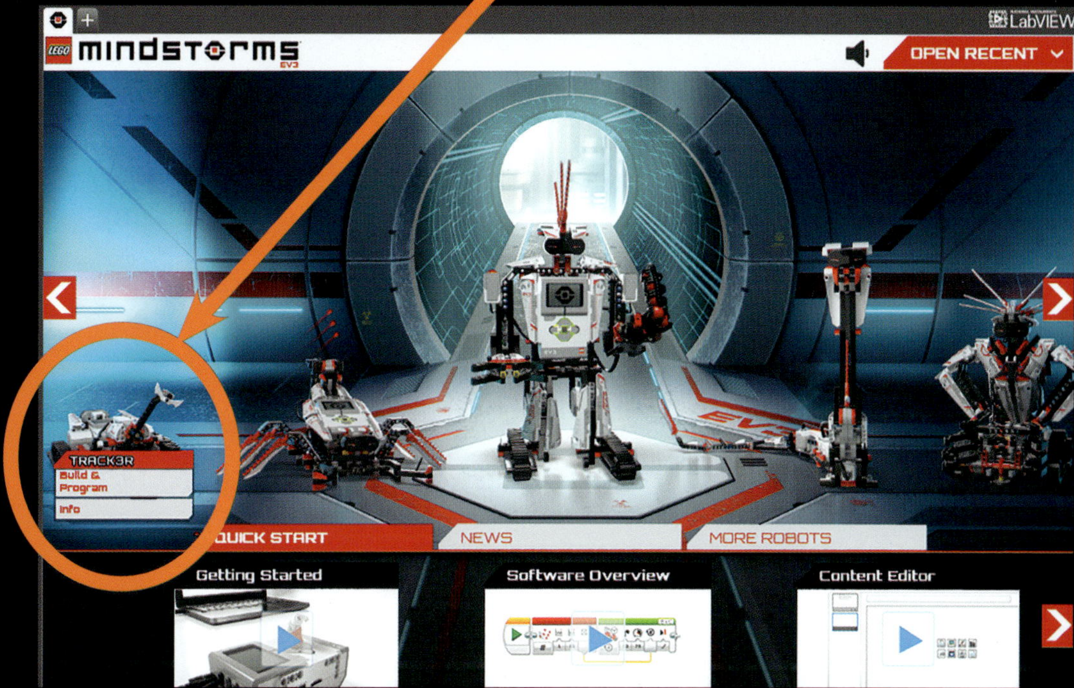

EV3 Operation

The EV3 brick is easy to use, and it can be connected to your robot or app via USB cable or **Bluetooth**. To turn on the EV3 brick, press the center button to bring up the start menu and a green light. Use the left and right buttons to navigate, or find your way through, the menu from there.

The MINDSTORMS Environment

Spend some time exploring the environment for the MINDSTORMS coding language. MINDSTORMS uses a graphical user interface, or GUI (GOO-ee). This is different from many traditional coding languages because it uses graphics, or pictures, instead of words to create a program.

Programming with MINDSTORMS is simple: Drag and drop programming blocks into the workspace, or canvas, to create **scripts**. The more blocks in the sequence, or list, the longer the code gets. All the blocks fit together like LEGO blocks. Make sure the first block in any sequence of your code is always a Start block.

This simple code shows a few common blocks you'll use when coding your MINDSTORMS robot. Notice the program begins with a Start block!

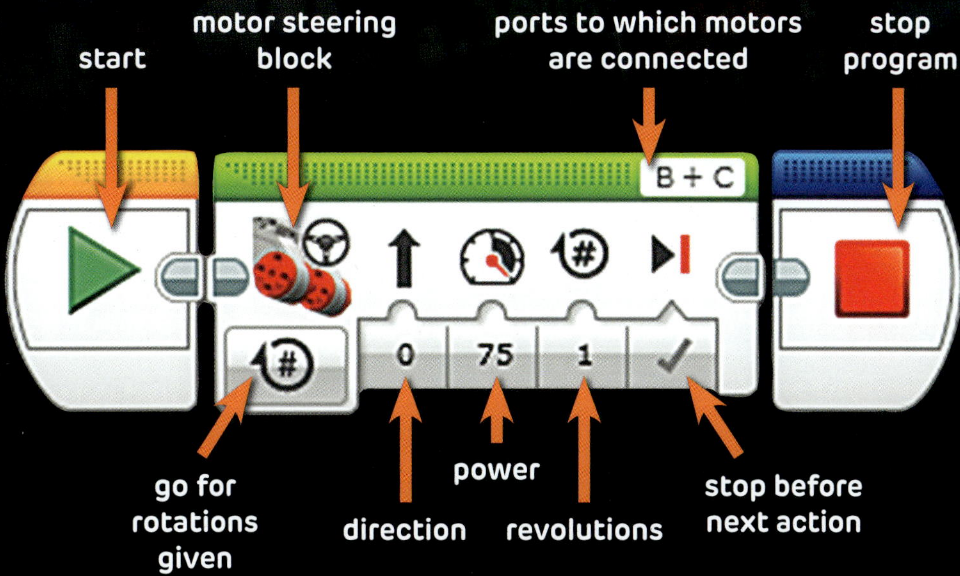

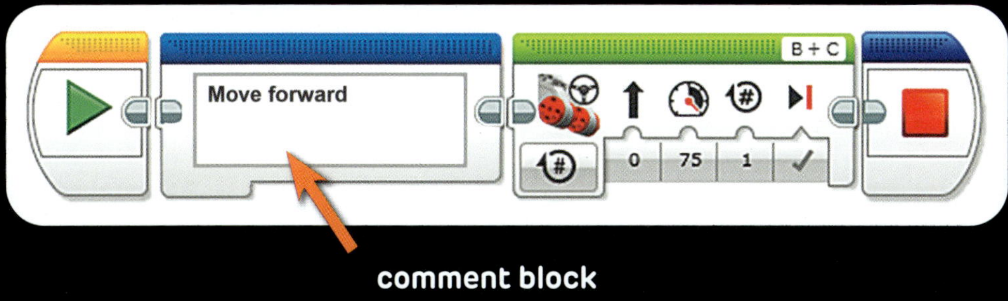

The Algorithm

An important block for beginners is the blue Comment block. These blocks don't affect your code. They serve as helpful notes to describe, or tell about, what each step in your code should be doing. You can use them to write a plan before you begin coding. This step-by-step plan is called an algorithm.

Making Connections

Data Wires are used for connecting blocks so you can create more advanced sequences. Data Wires allow the blocks to share information with each other even if they aren't right next to each other.

Data Wires also allow your robot to do two things at the same time. You can connect one Wire from the Start block to a Sound block and another one to a Movement block. When you start the program, the robot will move forward and play the sound at the same time. You need to add a Stop block to one of the blocks to turn off the robot once it completes the actions.

Keep It Organized

Data Wires work for Comment blocks too. You can keep Comments separate from your sequence but connect them to their relevant code blocks to keep things organized. Select a connecting point on a block to make a spool of wire appear. Click and drag the wire to the corresponding, or matching, block to link them.

This is the code to make your robot move and make a sound at the same time. Change the values in the blocks to get different results.

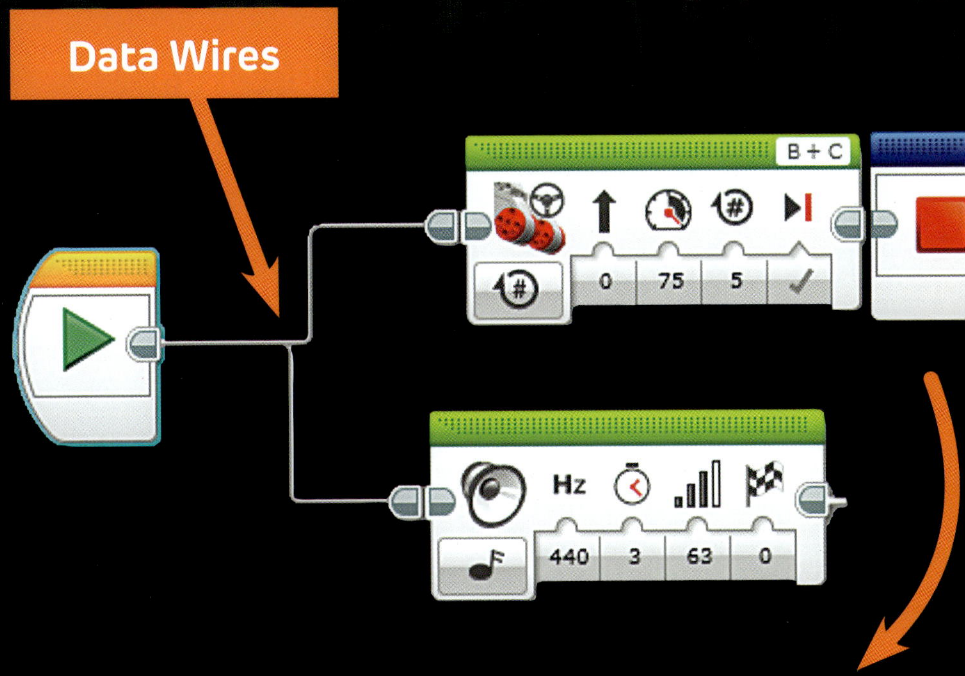

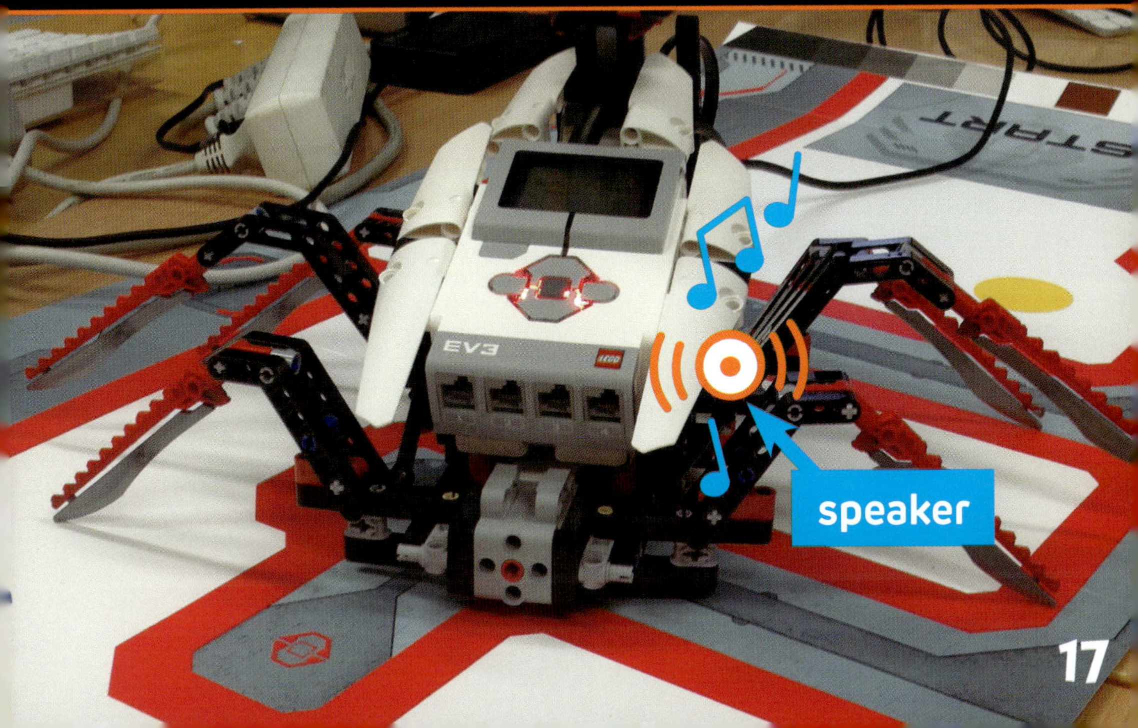

Around and Around

Another important block in your virtual toolkit is the Loop block. These orange-and-white blocks can hold other blocks inside them: if you put a sequence of blocks inside a Loop block, the program inside will repeat itself, or loop. The value you give to the Loop block indicates how many times the program will repeat—you can even input an infinity symbol, which means an unlimited amount of time!

Work Smarter, Not Harder

Programmers often reuse sections of code rather than typing them over and over again. This is smart coding. The Loop block gives you an idea of how this works, and you can bring this idea to any programming language you work with later.

Practice writing simple loops to see what Track3r is capable of. Don't forget to try adding sounds and different movements too!

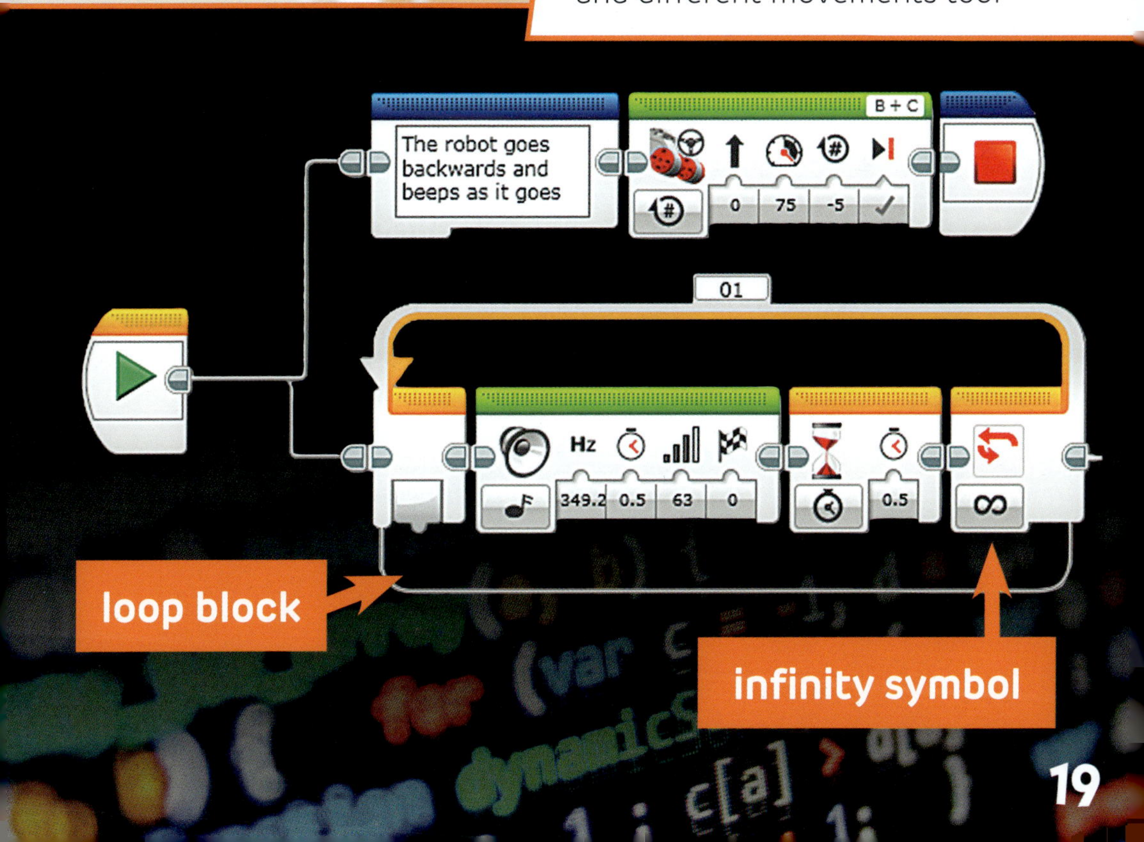

The Next Invention

The Robot Inventor set is a MINDSTORMS collection with its own style and different capabilities than the EV3 set. Programming robots from this set is a little different than coding the EV3 ones, but the Inventor set has many tutorials, or lessons, to help you get started.

With the Inventor set and app, you can code in a language that's a little different from the EV3's graphics-based language. It's similar to the programming language called Scratch. There are more block types, and the blocks connect from top to bottom instead of left to right.

The Inventor app comes with preprogrammed activities for every robot in the Inventor set. These activities let you experiment with existing code before writing your own.

From Blocks to Text

The first step of any project is to select the Word Block language or Python, which uses words instead of word blocks. Python is an excellent word-based language because it's more **concise** than other languages. In MINDSTORMS, you can even use the Word Block Translator tool to change different commands into Python and help you move from graphics to text, or words.

It's Alive!

Charlie is one of the robots you can program from the Inventor set. With Charlie, the Hub sits on its side and serves as Charlie's face. You can program the Hub's light sensors so they create different faces for Charlie. This kind of movement is called animation, and it makes Charlie's face very lifelike.

To do this, select Light from the block menu on the left side of the canvas, then choose a Start Animation block. Select the drop-down menu on the Start Animation block to get to the animation editor and library. You can create your own animation with the editor, use a preset animation from the library, or edit a preset animation to put your own spin on things.

CHARLIE

Wake Charlie up to see what it can do—one of Charlie's animations is even called "Wake Up!"

Hello, World!

A "Hello, World!" program is a coding exercise for beginners. With text-based languages, coders write a program to say "Hello, World!" with words. With MINDSTORMS, this exercise can entail writing a simple program to animate the robot's light sensors so it looks like the robot is waking up to say hello.

On the next page, you'll see a simple coding exercise for animating Charlie's light sensors. In this example, you'll use blocks from the Events, Light, and Control groups on the left of your canvas. Just as with any other program you write, make sure you add a Start block to the beginning of the sequence.

Choose the "Cool" preset animation from the animation library for this exercise. You can use it as is or change it in the editor. You can preview the animation by clicking the "Play" button in the bottom right corner of the editor. Complete your code with a Stop block.

Pixel by Pixel

The squares that light up on the Hub's display surface are pixels. "Pixel" is short for "picture element," and it's the smallest unit you'll see on any digital display. Usually pixels are incredibly small, with millions of single-color units making up an image. The Hub's pixels are much bigger, making them easy to work with.

What kind of personality would you program for Charlie? Have fun writing different code!

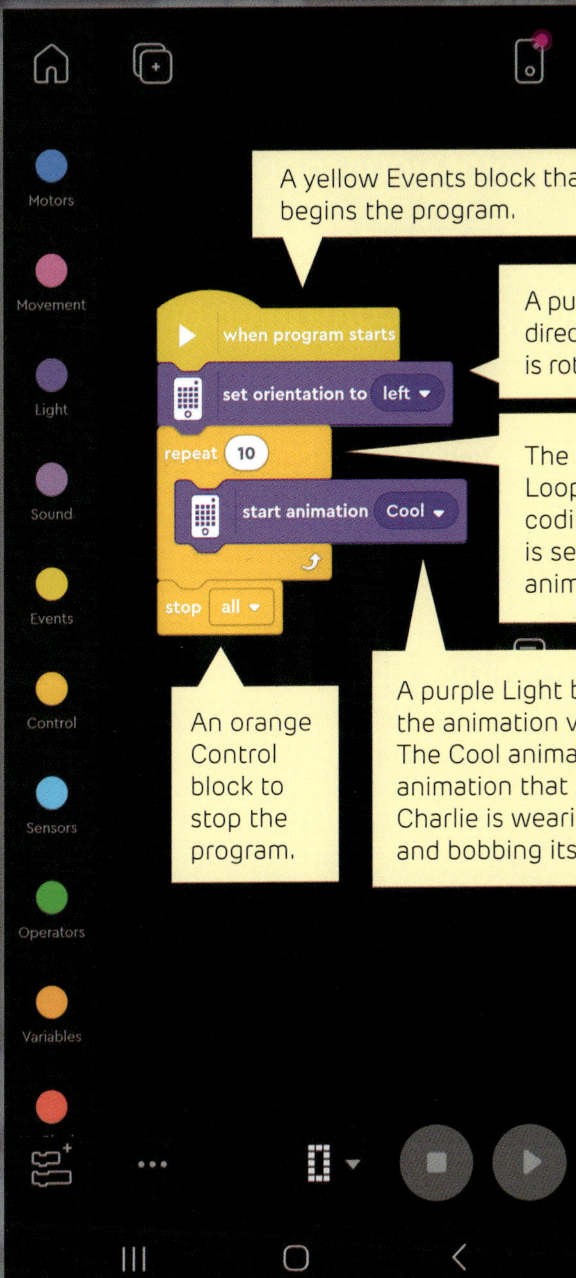

A yellow Events block that begins the program.

A purple Light block that changes the direction of the Hub. The Hub display is rotated to function as Charlie's face.

The Loop block is back! Like the EV3 Loop block, this block holds other coding blocks inside it. This block is set to 10 loops, which means the animation will play 10 times.

An orange Control block to stop the program.

A purple Light block where the animation value is selected. The Cool animation is a preset animation that makes it look like Charlie is wearing sunglasses and bobbing its head to a beat.

Virtual Remote

With the Inventor set, you can also program a controller so you can pilot your robot in real time. On the right side of your canvas, select the icon that looks like a video game controller, or gamepad. This will bring you to a separate canvas where you'll **configure** the controls you want to add.

Return to the regular programming canvas and you'll find a new block at the bottom of your block menu called Remote Control. Now you can write code for the controller buttons you've added. This turns your device into a controller without any extra equipment needed.

Machine Learning

Another way to control an Inventor robot is with machine learning. With machine learning, you give a program a set of examples to learn from. If you give the robot a few examples of you saying "start" into the microphone, that will help the robot learn how to recognize when you say "start" during regular use.

You can use your device as the controller or even connect a PlayStation 4 or Xbox One controller!

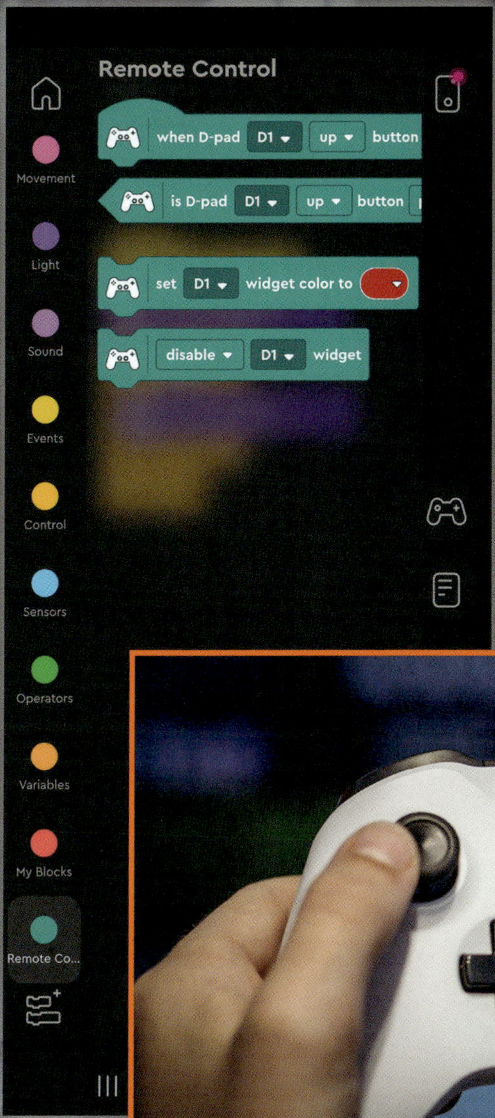

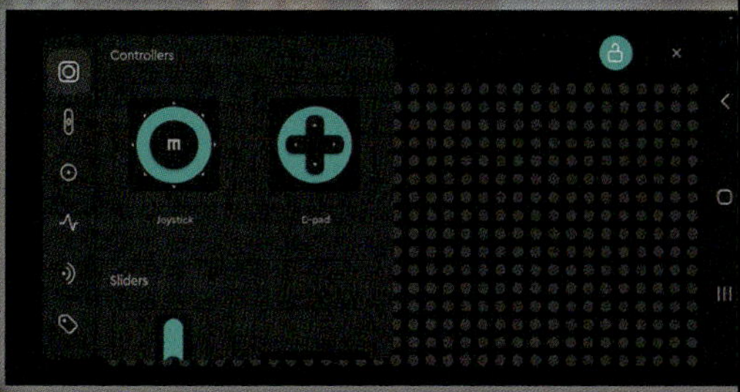

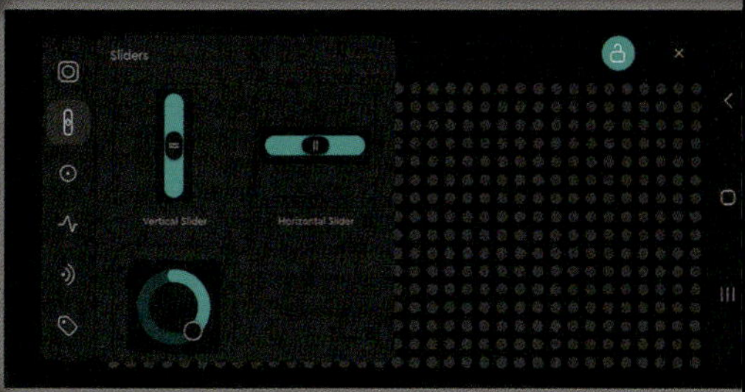

Coding Today and Tomorrow

One of the unique things about coding with LEGO MINDSTORMS is that you can see your code function in the real world right away. Playing with something **tangible** can help you learn, and MINDSTORMS robots are great for letting you play in both the real and virtual world.

Sharing the building and coding process with others can help you learn and have more fun too. Did you know there are robotics leagues you can join? *FIRST* LEGO League, for example, uses LEGO robots to teach kids coding and robotics on a global scale. If you love coding with MINDSTORMS today, think of what you might code tomorrow!

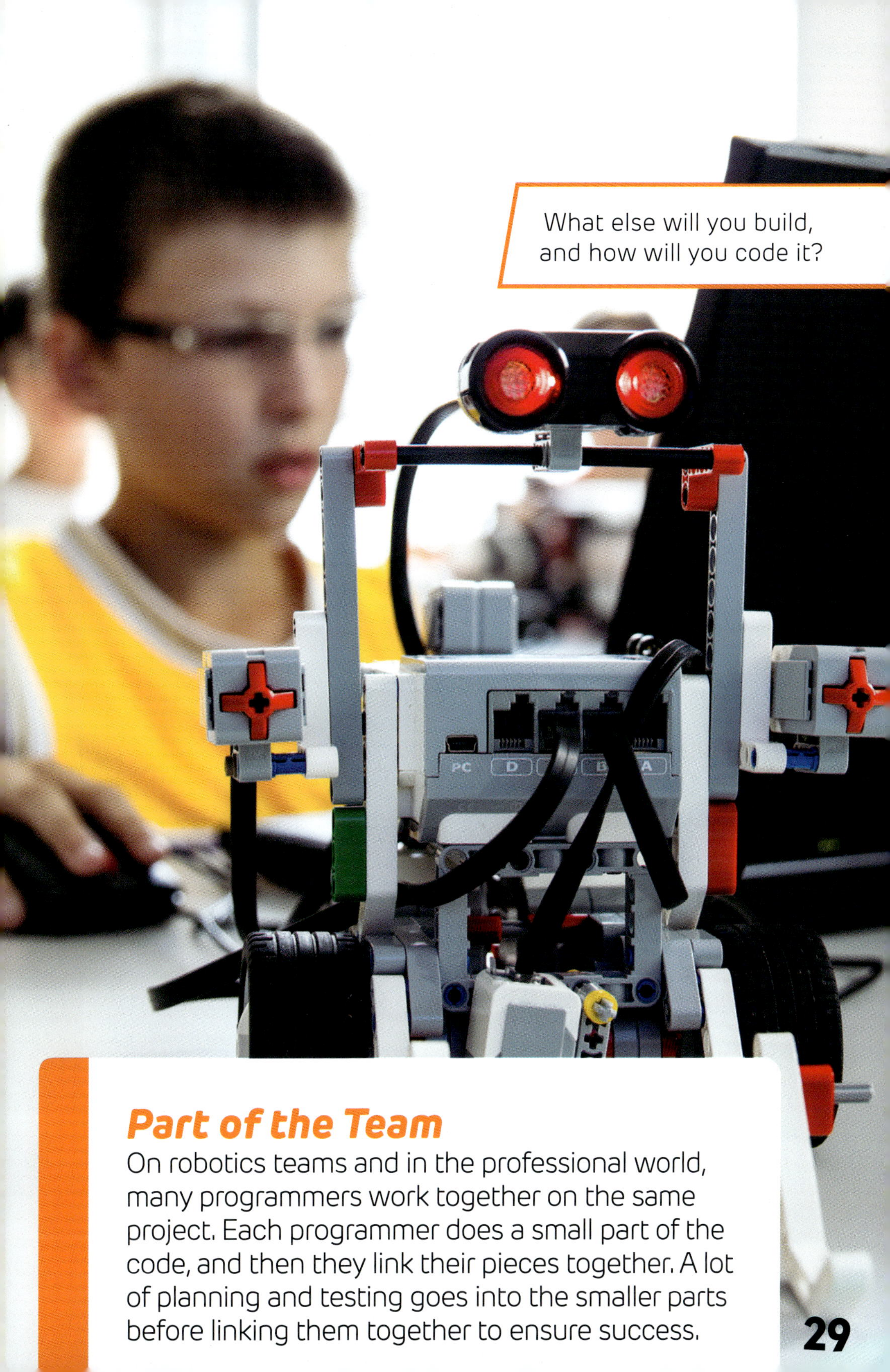

What else will you build, and how will you code it?

Part of the Team

On robotics teams and in the professional world, many programmers work together on the same project. Each programmer does a small part of the code, and then they link their pieces together. A lot of planning and testing goes into the smaller parts before linking them together to ensure success.

GLOSSARY

application: A software program, usually for a mobile device, that serves a specific purpose.

Bluetooth: A standard for connecting devices within a short range of each other wirelessly.

compatible: Capable of being used with another system or device without alteration.

concise: Brief but containing all necessary information.

configure: To arrange hardware and software for a specific purpose.

environment: The combination of computer hardware and software that allows a user to perform various tasks.

logic: A proper or reasonable way of thinking about or understanding something.

script: A sequence of instructions carried out by a computer program.

sensor: A device that reacts to a change in the world around it and then performs a task.

software: A program that runs on a computer and performs certain tasks.

tangible: Touchable; real in a substantial, material way.

vehicle: A machine that carries or transports something.

virtual: Occurring through a computer.

FOR MORE INFORMATION

Books

Bratzel, Barbara and Rob Torok. *Mastering LEGO MINDSTORMS: Build Better Robots with Python and Word Blocks*. San Francisco, CA: No Starch Press, 2022.

Maurer, Aaron. *Smart Robotics with LEGO MINDSTORMS Robot Inventor*. Birmingham, UK: Packt Publishing, 2021.

McCue, Camille. *Getting Started with Coding: Get Creative with Code!* Hoboken, NJ: Wiley, 2020.

Websites

Code.org
www.code.org
Providing computer science tools for students in kindergarten through 12th grade.

***FIRST* LEGO League**
www.firstinspires.org
Brings children ages 4 to 16 together to learn, build, code, and compete with LEGO robots.

LEGO
www.lego.com
The LEGO Group's official website.

Publisher's note to educators and parents: *Our editors have carefully reviewed these websites to ensure that they are suitable for students. Many websites change frequently, however, and we cannot guarantee that a site's future contents will continue to meet our high standards of quality and educational value. Be advised that students should be closely supervised whenever they access the internet.*

INDEX

algorithm, 15
application, 10, 11, 12, 13, 20, 21
battery, 6
Bluetooth, 13
Charlie, 22, 23, 24, 25
controller, 26, 27
Data Wire, 16, 17
EV3 set, 6, 10, 11, 12, 13, 20, 25
graphical user interface (GUI), 14
hardware, 8
Hub, 10, 11, 22, 24, 25
internet, 4
motor, 6, 7
Python, 21
Robot Inventor set, 10, 20, 21, 22, 26
Scratch, 20
script, 14
sensor, 6, 7, 11
software, 8, 10
Technic set, 6
Track3r, 9, 10, 12, 13, 19
USB, 13